# Future in Your Hands
## Building Smart Machines with Machine Learning Principles

## BY
Legend Salgado

# CONTENTS

INTRODUCTION ................................................................. 1

**Part 1:** ............................................................................. 8
**Foundations of Machine Learning** ............................... 8

Chapter 01 ........................................................................ 9
The Building Blocks of Machine Learning ....................... 9
Chapter 02 ...................................................................... 17
Getting Your Hands Dirty with Data .............................. 17

**Part 2:** ........................................................................... 27
**Building and Training Models** ..................................... 27

Chapter 03 ...................................................................... 28
Selecting the Right Machine Learning Model ................ 28
Chapter 04 ...................................................................... 36
Training Your First Model .............................................. 36

**Part 3:** ........................................................................... 46
**Advanced Techniques and Real-World Applications** .. 46

Chapter 05 ...................................................................... 47
Deep Dive into Neural Networks ................................... 47
Chapter 06 ...................................................................... 57
Machine Learning in Real-World Applications .............. 57

**Part 4** ............................................................................ 66
**Innovating with Machine Learning** ............................. 66

Chapter 07 ...................................................................... 67
Exploring Unsupervised Learning .................................. 67

..............................................................................67
Chapter 08.................................................................77
Reinforcement Learning and Future Trends....................77

**Conclusion** ................................................................**88**

# INTRODUCTION

# Understanding the Journey Ahead

Welcome to *The Future in Your Hands: Building Smart Machines with Machine Learning Principles*. This guidebook is designed to take you on an exciting journey into the world of machine learning (ML), a transformative field that's reshaping industries and unlocking new possibilities every day. Whether you're a student, an innovator, or simply curious about the technology behind smart machines, this book will provide the step-by-step guidance you need to understand, build, and apply ML in meaningful ways.

## What Is Machine Learning?

At its core, machine learning is the science of teaching computers to learn from data without being explicitly programmed. Let's break this down with a simple analogy:
Imagine teaching a child how to recognize dogs. Instead of describing every possible feature of a dog, you'd show the child pictures of dogs and say, "This is a dog." Over time, the child learns to identify dogs based on patterns—like their shape, size, or fur—without needing explicit rules. Machine learning works similarly, except the "child" is a computer, and the data serves as the teacher.

## Key Components of Machine Learning

1. **Data:** The foundation of ML. Computers learn from examples provided in the form of datasets.
2. **Algorithms:** These are like the brains of ML, helping systems find patterns in data.

3. **Models:** A model is the result of training an algorithm on data. It's the machine's way of understanding what it has learned.
4. **Feedback:** Just like humans improve with practice, machines refine their models through testing and feedback.

## Traditional Programming vs. Machine Learning

In traditional programming, humans write explicit instructions for computers to follow. For example, a program might calculate sales tax by multiplying a price by a predefined rate. In machine learning, instead of writing these explicit instructions, we provide data and let the machine determine patterns or rules on its own.

| Traditional Programming | Machine Learning |
|---|---|
| Rules and logic are predefined. | Rules are derived from the data. |
| Focuses on step-by-step tasks. | Focuses on adaptability and learning. |
| Limited by programmer knowledge. | Can discover new patterns autonomously. |

This adaptability makes ML especially powerful for solving complex problems where explicit rules are hard to define, like recognizing faces in photos or predicting stock prices.

## Why Machine Learning Matters Today

Machine learning isn't just a buzzword—it's a technology with real-world impact, powering innovations across industries. Here's a look at how ML is transforming our world:

## Healthcare

- ML is revolutionizing diagnostics by analyzing medical images to detect diseases like cancer early.
- Predictive models help doctors personalize treatments based on a patient's history and genetics.
- AI-powered tools are assisting in drug discovery, reducing time and costs.

## Finance

- Fraud detection systems use ML to identify unusual patterns in transactions, preventing billions in losses annually.
- Algorithms predict stock market trends by analyzing vast amounts of financial data in real time.
- Banks use ML to automate loan approvals and assess creditworthiness.

## Entertainment

- Streaming platforms like Netflix and Spotify use ML to recommend shows and music tailored to individual tastes.
- Video game developers employ ML to create smarter, more adaptive in-game characters.

## Retail

- E-commerce sites use ML to analyze shopping habits and offer personalized recommendations.
- Dynamic pricing strategies adjust product prices in real time based on demand and competition.

## How ML Is Shaping the Future

From self-driving cars to language translation apps, ML is enabling technologies once thought to exist only in science fiction. The future promises even more breakthroughs, like AI-powered robotics, improved climate predictions, and tools to combat misinformation.

## Who Should Read This Book?

This book is for anyone curious about machine learning, regardless of background.

### For Students

Are you new to machine learning and unsure where to start? This guide will break down concepts into digestible steps, empowering you to experiment with ML projects by the time you finish.

### For Innovators

If you're an entrepreneur or problem-solver, this book will equip you with the tools to harness ML for building smarter solutions, whether it's automating processes, analyzing data, or developing new products.

### For the Simply Curious

Even if you have no technical experience, this guide is written in an accessible, friendly tone to help you grasp the potential of ML and spark ideas for its application.

The only prerequisites are curiosity, determination, and an eagerness to learn.

## How to Use This Guide

This book is structured to take you step-by-step from foundational concepts to building and applying ML models. Here's how to navigate it:

## Overview of Parts and Chapters

- **Part I: Foundations of Machine Learning**
  Understand the basics of data, algorithms, and the ML workflow. Perfect for beginners.
- **Part II: Building and Training Models**
  Learn how to prepare data, select algorithms, and create your first ML models.
- **Part III: Advanced Techniques and Real-World Applications**
  Dive into neural networks, explore cutting-edge applications, and tackle real-world challenges.
- **Part IV: Innovating with Machine Learning**
  Explore unsupervised learning, reinforcement learning, and future trends in AI.

## Step-by-Step Roadmap

1. Start with foundational concepts to build a strong understanding of how ML works.

2. Progress to practical hands-on tasks like training models and analyzing data.
3. Experiment with advanced techniques and explore the possibilities of ML in different industries.
4. Apply your skills through real-world projects and innovate using ML principles.

## Encouragement Along the Way

Learning ML can seem daunting at first, but remember: every expert started as a beginner. This book is designed to be your companion, providing clear explanations, actionable steps, and motivating examples to keep you inspired.

By the end of this guide, you'll not only understand the principles of machine learning but also be equipped to build smart machines and shape the future of innovation.

Let's begin this exciting journey together!

# Part 1: Foundations of Machine Learning

# Chapter 01

## The Building Blocks of Machine Learning

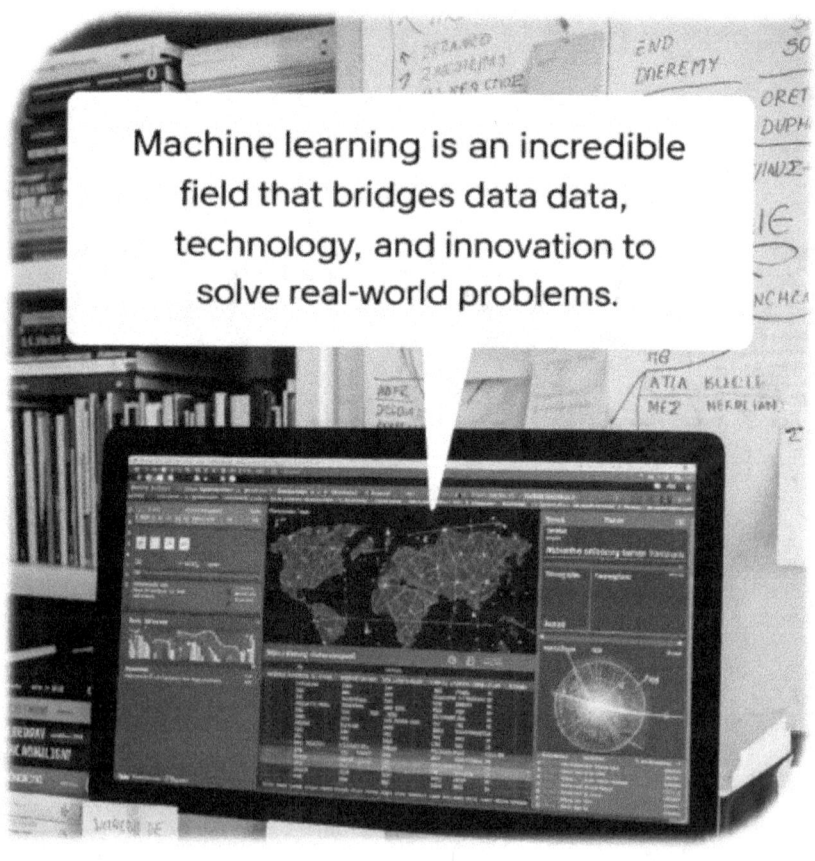

Machine learning is an incredible field that bridges data data, technology, and innovation to solve real-world problems.

Machine learning (ML) is an incredible field that bridges data, technology, and innovation to solve real-world problems. In this chapter, we'll break down the fundamental building blocks of ML, equipping you with the foundational knowledge needed to start your journey.

## The Basics of Algorithms and Data

Machine learning thrives on data. Without data, there is no learning. But it's not just about having data—it's about understanding how algorithms use data to find patterns and make decisions.

## The Role of Algorithms in Machine Learning

Algorithms in ML act like chefs following recipes. Given the right ingredients (data) and instructions (algorithm), the outcome is a well-prepared model capable of making predictions or uncovering insights.

## Three Core Types of Learning

ML is broadly categorized into three types of learning based on the data and the goals:

## Supervised Learning

- **Definition:** The algorithm learns from labeled data, where each input comes with an associated output.
- **Examples:** Predicting housing prices based on size (input) and past sale prices (output).

- **Common Algorithms:** Linear regression, support vector machines, neural networks.

## Unsupervised Learning

- **Definition:** The algorithm explores data without predefined labels, seeking hidden structures or patterns.
- **Examples:** Grouping customers by purchasing habits (clustering).
- **Common Algorithms:** K-means, hierarchical clustering, principal component analysis (PCA).

## Reinforcement Learning

1. **Definition:** The algorithm learns by interacting with an environment and receiving rewards or penalties.
2. **Examples:** Training a robot to navigate a maze or teaching AI to play chess.
3. **Key Feature:** Trial and error approach with a focus on maximizing long-term rewards.

### Key Terminologies You Need to Know

To navigate the ML world, you need to familiarize yourself with essential terms. These are the building blocks of your ML vocabulary:

- **Features:** The inputs to the model. For example, in predicting housing prices, features could include the size, location, and number of bedrooms.

- **Labels:** The outputs or results associated with features, such as the price of a house.
- **Datasets:** Collections of data used for training, testing, and validating a model.
- **Models:** The mathematical representation of a process, trained to recognize patterns in data.
- **Training and Testing:**

- ✓ **Training Data:** Data used to teach the model.
- ✓ **Testing Data:** Data used to evaluate the model's accuracy and performance.

## Types of Problems Machine Learning Solves

Machine learning isn't one-size-fits-all—it's a versatile tool suited to different kinds of problems:

### Classification

- **Definition:** Assigning labels to inputs.
- **Example:** Determining whether an email is spam or not.
- **Real-World Use:** Fraud detection, medical diagnostics.

### Regression

- ✓ **Definition:** Predicting continuous numerical values.
- ✓ **Example:** Estimating stock prices based on historical data.

- ✓ **Real-World Use:** Weather forecasting, financial predictions.

## Clustering

- ✓ **Definition:** Grouping similar items without predefined labels.
- ✓ **Example:** Categorizing users based on browsing behavior.
- ✓ **Real-World Use:** Customer segmentation, anomaly detection.

## Recommendation Systems

- ➢ **Definition:** Suggesting items based on user preferences.
- ➢ **Example:** Netflix recommending movies.
- ➢ **Real-World Use:** E-commerce, media platforms.

## Breaking Down the Machine Learning Workflow

Understanding the ML workflow helps you visualize how data transforms into actionable insights. Here's a step-by-step breakdown:

### 1. Data Collection

- ➢ Identify and gather relevant data from reliable sources (e.g., databases, APIs).

### 2. Data Preparation

- Clean the data: Remove duplicates, handle missing values, and eliminate errors.
- Normalize and preprocess: Ensure data is in a usable format.

## 3. Training the Model

- Feed the training dataset to the algorithm to learn patterns.

## 4. Testing and Validation

- Test the model on unseen data to evaluate performance.

## 5. Deployment

- Implement the model into real-world systems for continuous learning and feedback.

## Examples of Everyday Machine Learning

ML may seem like a complex concept, but you encounter it every day, often without realizing it:

### Personalized Recommendations

- Streaming services like Netflix and Spotify suggest content based on your viewing or listening history.

### Spam Detection

- Email services use ML to filter out unwanted messages by recognizing patterns in spam emails.

## Image Recognition

➢ Applications like Google Photos use ML to identify objects and faces in your pictures.

## Virtual Assistants

➢ Tools like Siri and Alexa rely on ML for natural language processing and voice recognition.

# Common Misconceptions About ML

Many myths surround machine learning, and debunking them is crucial for setting realistic expectations:

**Myth:** ML can solve all problems.

➢ **Reality:** ML is powerful but not magical. It depends on data quality, quantity, and the problem domain.

**Myth:** You need to be a math genius to understand ML.

➢ **Reality:** While math is important, tools and libraries make it accessible to those with basic programming knowledge.

**Myth:** ML models are 100% accurate.

➢ **Reality:** No model is perfect. The goal is to create models that are accurate enough to add value.

**Myth:** ML will replace humans.

- **Reality:** ML complements human skills, automating repetitive tasks while freeing humans for creative and strategic roles.

Machine learning is a fascinating journey that starts with understanding these core concepts. With the foundation laid, you're ready to dive deeper into the hands-on aspects of data and algorithms, exploring how to build, train, and evaluate models in practical applications. Let's continue!

# Chapter 02

## Getting Your Hands Dirty with Data

Data is the fuel that powers machine learning. Without quality data, even the most sophisticated algorithms fail to deliver meaningful results. In this chapter, you'll learn the fundamentals of working with datasets, from locating open-source data to cleaning, visualizing, and analyzing it. By the end, you'll have the skills to tackle your first dataset exploration project.

## Introduction to Datasets

Datasets are collections of data organized for analysis. They form the backbone of any machine learning project. Choosing the right dataset and understanding its structure are critical first steps.

## Where to Find Open-Source Data

Open-source datasets are freely available and serve as an excellent starting point for beginners. Here are a few popular sources:

## Kaggle

- A treasure trove of datasets across domains like healthcare, finance, and sports.
- Offers community support and competitions to test your skills.
- Visit: https://www.kaggle.com

## UCI Machine Learning Repository

- A classic collection of datasets for academic and research purposes.
- Visit: : https://archive.ics.uci.edu/ml

## Google Dataset Search

- A search engine for datasets across the web.
- Visit: https://datasetsearch.research.google.com

## Government Portals

- Examples include Data.gov (U.S.), Data.gov.uk (U.K.), and European Data Portal.

## How to Choose the Right Dataset

When selecting a dataset for your project, consider the following factors:

1. **Relevance:** Does the dataset align with your project's goals?
2. **Size:** Is the dataset large enough to train your model effectively?
3. **Quality:** Are there significant errors, missing values, or inconsistencies?
4. **Diversity:** Does the data capture a wide range of scenarios or conditions?

## Understanding Data Formats

Data comes in various formats, and understanding them is essential for effective handling.

### 1. CSV (Comma-Separated Values)

- **Structure:** Tabular data stored as plain text, with each line representing a record.

> **Usage:** Ideal for structured datasets like sales records or survey results.

## Example:

```
Copy code
Name, Age, Country
Alice, 30, USA
Bob, 25, UK
```

## 2. JSON (JavaScript Object Notation)

> **Structure:** Hierarchical, representing data as key-value pairs.
> **Usage:** Suitable for complex, nested data like APIs or web scraping results.

## Example:

```json
Copy code
{
    "Name": "Alice",
    "Age": 30,
    "Country": "USA"}
```

## 3. Image Datasets

> **Structure:** Collections of image files, often accompanied by labels in a separate file.
> **Usage:** Common in computer vision tasks like object detection or face recognition.

- **Example:** A folder of cat and dog images with labels indicating their categories.

## Cleaning and Preprocessing Data

Raw data is rarely perfect. Before feeding it into a machine learning model, you need to clean and preprocess it.

## Steps in Data Cleaning

### Handle Missing Values:

- Replace missing values with mean/median for numerical data or mode for categorical data.
- Alternatively, remove rows or columns with excessive missing values.

### Remove Outliers:

- Identify and eliminate outliers using statistical methods like Z-scores or visualization tools.

### Fix Inconsistencies:

- Standardize formats (e.g., date formats or text casing).
- Correct errors like duplicate records or misspelled entries.

### Normalize or Scale Data:

- For numerical data, ensure values fall within a comparable range using techniques like Min-Max Scaling or Standardization.

# Introduction to Python for Data Handling

Python's libraries make data handling efficient and accessible. Two essential libraries for this task are Pandas and NumPy.

## 1. Pandas Basics

### Loading Data:

```python
Copy code
import pandas as pd
    data = pd.read_csv("file.csv")
```

### Exploring Data:

```python
Copy code
print(data.head())   # View the first few rowsprint(data.info())   # Check data types and missing values
```

### Cleaning Data:

```python
Copy code
data = data.dropna()   # Remove missing values
data['column'] = data['column'].fillna(data['column'].mean())   # Fill missing values
```

## 2. NumPy Basics

## Creating Arrays:

```python
Copy code
import numpy as np
    array = np.array([1, 2, 3, 4])
```

## Performing Operations:

```python
Copy code
print(np.mean(array))    # Calculate
meanprint(np.sum(array))    # Calculate sum
```

## Data Visualization Techniques

Data visualization is a powerful way to uncover trends and patterns. Libraries like Matplotlib and Seaborn make visualization easy and intuitive.

## Using Matplotlib

### Line Plot:

```python
Copy code
import matplotlib.pyplot as plt
    plt.plot([1, 2, 3], [4, 5, 6])
    plt.show()
```

### Bar Chart:

```python
```

```
Copy code
plt.bar(['A', 'B', 'C'], [10, 15, 7])
     plt.show()
```

## Using Seaborn

## Correlation Heatmap:

```python
Copy code
import seaborn as sns
sns.heatmap(data.corr(), annot=True)
plt.show()
```

## Scatter Plot:

```python
Copy code
sns.scatterplot(x='feature1', y='feature2', data=data)
plt.show()
```

## Action Plan: Your First Dataset Exploration Project

It's time to apply what you've learned with a simple hands-on project.

### Steps to Explore a Dataset

## Choose a Dataset:

➢ Download the Titanic dataset from Kaggle.

## Load the Data:

➢ Use Pandas to load and inspect the dataset.

```python
Copy code
data = pd.read_csv("titanic.csv")
print(data.head())
```

## Clean the Data:

➢ Handle missing values in the Age column:

```python
Copy code
data['Age'] = data['Age'].fillna(data['Age'].mean())
```

## Visualize the Data:

➢ Create a bar chart of passenger survival rates:

```python
Copy code
sns.barplot(x='Survived', y='PassengerId', data=data)
plt.show()
```

## Analyze Trends:

➢ Identify trends in survival rates based on gender or class.

By completing this project, you'll gain practical experience in dataset handling, cleaning, and visualization. This foundational knowledge sets the stage for building and training machine learning models in the next chapters. Let's move forward!

# Part 2: Building and Training Models

# Chapter 03

## Selecting the Right Machine Learning Model

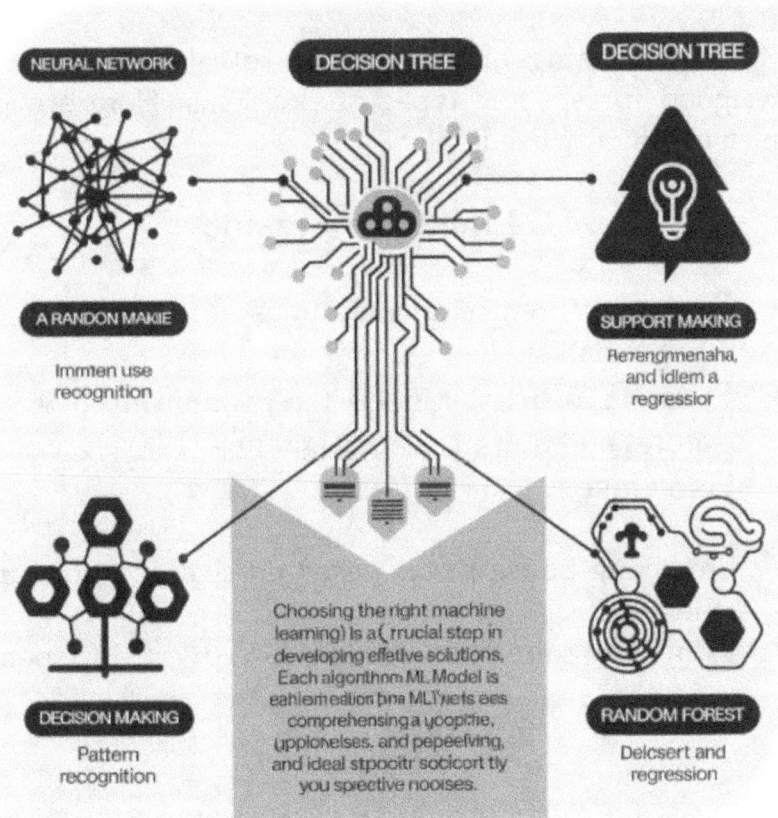

Choosing the right machine learning (ML) model is a crucial step in developing effective solutions. Each algorithm has its strengths, weaknesses, and ideal applications. This chapter provides a comprehensive guide to understanding, comparing, and selecting the appropriate model for your specific needs.

## Overview of Common Algorithms

Machine learning offers a variety of algorithms, each designed for specific types of problems. Here are the most commonly used ones:

### 1. Linear Regression

- **Purpose:** Predicts a continuous outcome based on input variables.
- **How It Works:** Models the relationship between features and output as a straight line.
- **Use Cases:**

✓ Predicting house prices based on size, location, and amenities.
✓ Estimating sales revenue based on advertising spend.

### 2. Decision Trees

- **Purpose:** Makes predictions by splitting data into branches based on feature values.
- **How It Works:** Forms a tree-like structure where each node represents a decision based on a feature.
- **Use Cases:**

- ✓ Loan approval systems.
- ✓ Identifying high-risk patients in healthcare.

### 3. Support Vector Machines (SVMs)

- ➤ **Purpose:** Classifies data points by finding the best boundary between categories.
- ➤ **How It Works:** Maximizes the margin between classes using hyperplanes.
- ➤ **Use Cases:**

- ✓ Image classification.
- ✓ Spam detection.

### 4. Neural Networks

- ➤ **Purpose:** Mimics the human brain to recognize complex patterns.
- ➤ **How It Works:** Uses layers of interconnected nodes to process data and make predictions.
- ➤ **Use Cases:**

- ✓ Facial recognition systems.
- ✓ Predicting stock market trends.

## When to Use Which Algorithm

Selecting the right algorithm depends on your data, problem type, and goals.

### 1. For Regression Problems

- ➤ **Use:** Linear Regression, Random Forest Regressor.

➤ **Examples:** Forecasting energy consumption or predicting customer lifetime value.

## 2. For Classification Problems

➤ **Use:** SVMs, Logistic Regression, Decision Trees.
➤ **Examples:** Classifying emails as spam or not, diagnosing diseases.

## 3. For Clustering or Grouping

➤ **Use:** K-Means, DBSCAN.
➤ **Examples:** Customer segmentation, anomaly detection.

## 4. For Sequential or Time-Series Data

➤ **Use:** Recurrent Neural Networks (RNNs), Long Short-Term Memory (LSTM).
➤ **Examples:** Predicting stock prices, weather forecasting.

## Exploring Trade-offs: Accuracy vs. Complexity

No algorithm is perfect. Every choice involves trade-offs between performance, computational cost, and interpretability.

### 1. Accuracy vs. Interpretability

- **Simple Models (e.g., Linear Regression):** Easier to interpret but may lack accuracy for complex data.
- **Complex Models (e.g., Neural Networks):** High accuracy for intricate patterns but difficult to explain.

### 2. Accuracy vs. Training Time

- **Decision Trees:** Quick to train but may not perform well on large datasets.
- **Neural Networks:** Accurate for large datasets but require substantial time and computational resources.

### 3. Overfitting vs. Underfitting

- **Overfitting:** Model learns noise in the training data, reducing its ability to generalize.
- **Underfitting:** Model is too simple, failing to capture the underlying trends.

## Step-by-Step: Choosing an Algorithm for Your Project

Follow these steps to identify the best algorithm:

## Define the Problem Type

- Is it classification, regression, or clustering?

## Evaluate the Data

- Size of the dataset, quality, and type (numerical, categorical, images).

## Consider Model Complexity

- For small datasets, start with simple models like linear regression.
- For large, complex datasets, explore advanced models like neural networks.

## Test Multiple Models

- Train and compare several algorithms on your data.
- Use performance metrics like accuracy, precision, and recall.

## Iterate and Optimize

- Adjust parameters to fine-tune the chosen model for better results.

## Practical Challenges in Model Selection

Even with a systematic approach, model selection comes with challenges:

### 1. Overfitting and Underfitting

- **Overfitting Fixes:** Use techniques like regularization (L1, L2) or pruning decision trees.
- **Underfitting Fixes:** Try more complex models or include additional features.

### 2. Data Limitations

> If data is limited, consider augmenting it through synthetic data generation or transfer learning.

## 3. Computational Constraints

> For resource-intensive models, use cloud-based services or optimized libraries like TensorFlow or PyTorch.

## Case Study: A Predictive Analytics Example

Let's walk through a practical example of choosing the right algorithm.

## Scenario: Predicting Student Performance

A school wants to predict whether a student will pass or fail based on their attendance, test scores, and participation.

## Step-by-Step Approach

### Define the Problem Type

> This is a classification problem (pass or fail).

### Analyze the Data

> Features: Attendance rate, test scores, participation level.
> Labels: Pass (1) or Fail (0).

### Test Candidate Models

- **Logistic Regression:** Quick and interpretable.
- **Decision Trees:** Handles non-linear patterns well.
- **Neural Networks:** Could capture complex relationships if enough data is available.

## Evaluate Models

- Compare accuracy, precision, recall, and training time.

## Choose the Best Model

- Decision Trees might balance interpretability and performance for this small dataset.

By the end of this chapter, you should feel confident in your ability to evaluate and choose machine learning models based on your project requirements. The next chapter will dive into the mechanics of training and testing these models to ensure they deliver meaningful results.

# Chapter 04
## Training Your First Model

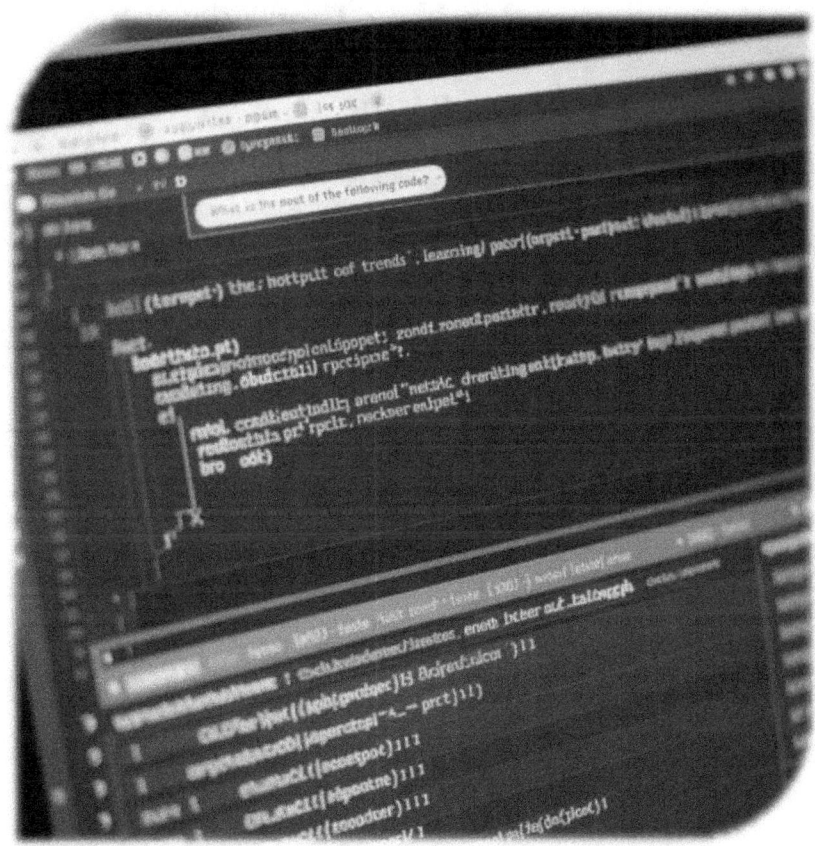

Training your first machine learning (ML) model is an exciting step in your journey. This chapter will guide you through setting up the tools, preparing data, building a simple model, and evaluating its performance. By the end, you'll complete your first ML project: predicting trends using linear regression.

## Setting Up Your Environment

Before you can train a model, you need a development environment equipped with the necessary tools and libraries.

## 1. Installing Jupyter Notebooks

Jupyter Notebooks is an interactive tool for writing and running Python code, perfect for ML experiments.

## Installation Steps:

3. Launch Jupyter:

```bash
jupyter notebook
```

1. Install Python (preferably version 3.8 or later).
2. Use pip to install Jupyter:

```bash
pip install notebook
```

Open your browser to the provided link (e.g., http://localhost:8888/) to access the notebook interface.

## Why Use Jupyter?

> Interactive: Allows you to execute code and see results instantly.
> Visual: Integrates graphs and charts seamlessly.
> Organized: Combines code, explanations, and visuals in one place.

## 2. Installing Essential Libraries

Python's ML ecosystem is robust and beginner-friendly. Install these key libraries:

- **NumPy** for numerical operations:

    ```bash
    pip install numpy
    ```

- **Pandas** for data manipulation:

    ```bash
    pip install pandas
    ```

- **Matplotlib** and **Seaborn** for visualization:

    ```bash
    pip install matplotlib seaborn
    ```

Scikit-Learn for ML algorithms:
bash
pip install scikit-learn

## Preparing Training and Testing Datasets

Machine learning models learn from data. Splitting your dataset into training and testing sets is crucial for reliable evaluation.

### 1. Why Split Data?

- **Training Set:** Used to train the model.
- **Testing Set:** Used to evaluate the model's performance on unseen data.
- Prevents overfitting by ensuring the model generalizes well to new data.

### 2. Splitting Data Effectively

Use Scikit-Learn's `train_test_split` function:

**from sklearn.model_selection import train_test_split**

**# Example dataset**
**X = data.drop(columns=['target'])  # Features**
**y = data['target']  # Labels**

**# Split into 80% training and 20% testing**
**X_train, X_test, y_train, y_test = train_test_split(X, y, test_size=0.2, random_state=42)**

# Building a Simple ML Model

Let's implement a simple linear regression model to predict trends. Linear regression is one of the most basic ML algorithms, making it ideal for beginners.

## 1. What is Linear Regression?

Linear regression predicts a continuous value (e.g., house prices, temperatures) based on input features by finding the best-fit line.

## 2. Step-by-Step Guide

### Import Libraries:

from sklearn.linear_model import LinearRegression

import pandas as pd

**Load Your Dataset:** Use a CSV file or create a small sample dataset:

data = pd.DataFrame({

    'feature1': [1, 2, 3, 4, 5],

    'feature2': [2, 4, 6, 8, 10],

    'target': [3, 6, 9, 12, 15]

})

## Prepare Features and Target:

```python
X = data[['feature1', 'feature2']]
y = data['target']
```

## Train the Model:

```python
model = LinearRegression()
model.fit(X, y)
```

## Make Predictions:

```python
Copy code
predictions = model.predict(X)
print("Predictions:", predictions)
```

# Hyperparameters and Their Tuning

Hyperparameters are settings you configure before training a model. Proper tuning can significantly improve performance.

## 1. Common Hyperparameters in Linear Regression

- **Fit Intercept:** Determines whether to calculate the y-intercept (default is `True`).
- **Normalize:** Normalizes the data before fitting (optional for scaled data).

## 2. Tuning Techniques

**Grid Search:** Test combinations of hyperparameters systematically:

```python
from sklearn.model_selection import GridSearchCV
    param_grid = {'normalize': [True, False]}
    grid_search = GridSearchCV(LinearRegression(),
    param_grid)
    grid_search.fit(X_train, y_train)
```

**Random Search:** Test random combinations for quicker results.

## Evaluating Model Accuracy

Evaluating your model ensures it meets the desired performance. Use metrics tailored to your problem type.

## 1. Regression Metrics

**Mean Absolute Error (MAE):** Average absolute error between predictions and actual values.

```python
from sklearn.metrics import mean_absolute_error
    mae = mean_absolute_error(y_test,
    predictions)print("Mean Absolute Error:", mae)
```

**Mean Squared Error (MSE):** Average squared error, penalizing larger errors more.

```python
```

```python
from sklearn.metrics import mean_squared_error
    mse = mean_squared_error(y_test, predictions)
print("Mean Squared Error:", mse)
```

**$R^2$ Score:** Measures how well the model explains variance in the data (1 is perfect).

```python
from sklearn.metrics import r2_score
    r2 = r2_score(y_test, predictions)
print("R² Score:", r2)
```

## 2. Visualizing Results

Plot predictions against actual values:

```python
Copy code
import matplotlib.pyplot as plt

plt.scatter(y_test, predictions)
plt.xlabel("Actual Values")
plt.ylabel("Predicted Values")
plt.title("Actual vs Predicted")
plt.show()
```

## Your First ML Project: Predicting Trends

### Project Goal

Predict housing prices based on features like square footage, number of bedrooms, and location.

### Step-by-Step Plan

## Select a Dataset:

Uspytho
```
from sklearn.datasets import fetch_california_housing
data = fetch_california_housing(as_frame=True)
df = data.frame
```

## Split the Data:

```python
X = df.drop(columns=['MedHouseVal'])  # Features
y = df['MedHouseVal']  # Target
X_train, X_test, y_train, y_test = train_test_split(X, y, test_size=0.2, random_state=42)
```

## Train a Model:

```python
model = LinearRegression()
model.fit(X_train, y_train)
```

## Evaluate the Model:

```python
predictions = model.predict(X_test)
print("MAE:", mean_absolute_error(y_test, predictions))
print("R² Score:", r2_score(y_test, predictions))
```

**Visualize Predictions:** Plot predicted vs actual housing prices.

Congratulations! You've trained and evaluated your first ML model. In the next chapter, we'll explore advanced techniques like model optimization and working with more complex datasets.

# Part 3: Advanced Techniques and Real-World Applications

# Chapter 05

## Deep Dive into Neural Networks

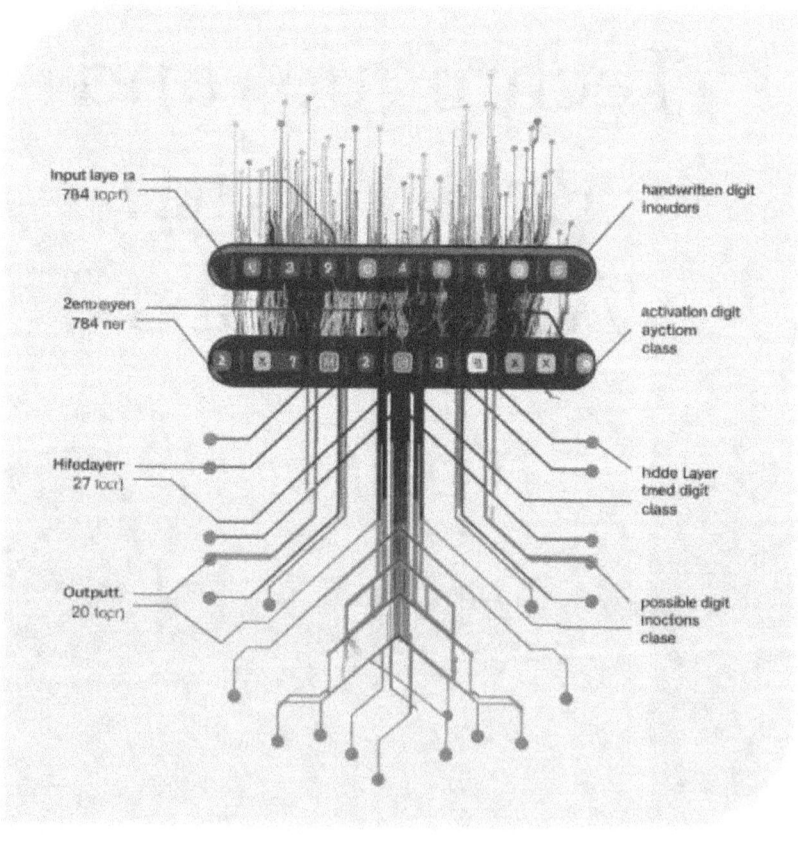

Neural networks, inspired by the human brain, are at the core of many advanced machine learning applications. From image recognition to natural language processing, these systems have revolutionized AI. In this chapter, we will explore the fundamentals of neural networks, their learning mechanisms, common architectures, and practical techniques to build and train your first network. We'll conclude with an exciting project: creating a digit recognizer using the MNIST dataset.

## What Are Neural Networks?

Neural networks are computational models that mimic the way neurons in the human brain work. They process input data through layers of interconnected nodes (neurons) to produce predictions or decisions.

### 1. Basics of Perceptrons

**Perceptron:** The simplest neural network unit that performs a linear classification.

1. **Input:** Takes features as numerical values (e.g., pixel brightness in images).
2. **Weights:** Assigns importance to each input.
3. **Activation Function:** Determines the output (e.g., 0 or 1 in classification problems).

#### How it works:

- Inputs are multiplied by weights and summed.
- The sum is passed through an activation function to produce the final output.

## 2. Hidden Layers

- Introduce complexity by adding intermediate layers between input and output.
- **Purpose:** Capture relationships that aren't linear, enabling the network to learn intricate patterns.

## 3. Activation Functions

- Add non-linearity, allowing neural networks to solve complex problems. Common types:

- ✓ **ReLU (Rectified Linear Unit):** Outputs the input directly if positive, else 0.
- ✓ **Sigmoid:** Converts values to a range between 0 and 1, often used in binary classification.
- ✓ **Softmax:** Converts outputs to probabilities, typically used in multi-class classification.

## How Neural Networks Learn

Neural networks learn through an iterative process of adjusting weights to minimize errors.

## 1. Backpropagation

- **Definition:** A method to calculate the error's impact on each weight.
- **Steps:**

1. Forward pass: Compute predictions using current weights.

2. Compute loss: Compare predictions with actual outputs using a loss function (e.g., Mean Squared Error for regression).
3. Backward pass: Use derivatives to update weights, reducing the loss.

## 2. Gradient Descent

- **Purpose:** An optimization technique to find the weights that minimize the loss function.
- **Variants:**
- **Batch Gradient Descent:** Uses the entire dataset to update weights.
- **Stochastic Gradient Descent (SGD):** Updates weights after each data point.
- **Mini-batch Gradient Descent:** A balance between batch and stochastic approaches.

## Common Architectures

Different neural network architectures are tailored for specific tasks.

## 1. Convolutional Neural Networks (CNNs)

**Purpose:** Designed for image processing.

**Key Features:**

- Convolution layers extract spatial features like edges and textures.
- Pooling layers reduce the dimensionality, retaining essential information.

**Applications:**

➢ Object detection, facial recognition, medical imaging.

## 2. Recurrent Neural Networks (RNNs)

**Purpose:** Handles sequential data like text and time series.

**Key Features:**

➢ Incorporates memory of previous inputs using loops in the network.
➢ Variants like LSTMs and GRUs address vanishing gradient issues.

**Applications:**

➢ Language translation, stock price prediction, speech recognition.

## 3. Transformers

**Purpose:** Revolutionize natural language processing (NLP) and large-scale data processing.

**Key Features:**

➢ Self-attention mechanism prioritizes relevant parts of input sequences.
➢ Handles long dependencies better than RNNs.

**Applications:**

➤ Chatbots, machine translation, document summarization.

## Exploring TensorFlow and PyTorch

TensorFlow and PyTorch are two of the most popular frameworks for building neural networks. Both provide user-friendly APIs for beginners and robust tools for advanced users.

### 1. Setting Up TensorFlowInstallation:

pip install tensorflow

```
Creating a Simple Neural Network

import tensorflow as tf

from tensorflow.keras import layers, models

# Define a simple feedforward network

model = models.Sequential([

    layers.Dense(128, activation='relu', input_shape=(784,)),

    layers.Dense(64, activation='relu'),

    layers.Dense(10, activation='softmax')

])
```

```python
# Compile the model
model.compile(optimizer='adam',
loss='categorical_crossentropy',
metrics=['accuracy'])
```

## 2. Setting Up PyTorch

### Installation

```
pip install torch torchvision
```

Creating a Simple Neural Network

```python
import torch
import torch.nn as nn

# Define a neural network
class SimpleNN(nn.Module):
    def __init__(self):
        super(SimpleNN, self).__init__()
        self.fc1 = nn.Linear(784, 128)
        self.fc2 = nn.Linear(128, 64)
        self.fc3 = nn.Linear(64, 10)
```

```
def forward(self, x):
    x = torch.relu(self.fc1(x))
    x = torch.relu(self.fc2(x))
    x = torch.softmax(self.fc3(x), dim=1)
    return x
```

## Avoiding Overfitting in Neural Networks

Overfitting occurs when a model learns noise in the training data, reducing its performance on new data.

### 1. Regularization Techniques

**L1 and L2 Regularization:** Add penalties to the loss function based on weight size.

**Dropout:** Randomly disables neurons during training, forcing the network to generalize.

### 2. Data Augmentation

➤ Generate additional training data by applying transformations like rotations, flips, and scaling.

### 3. Early Stopping

➤ Monitor validation performance and stop training when the model begins to overfit.

# Project Idea: Build a Digit Recognizer with MNIST Dataset

The MNIST dataset contains handwritten digits, a perfect starting point for applying neural networks.

## Steps to Build the Digit Recognizer:

### Load the Dataset:

```python
from tensorflow.keras.datasets import mnist
(X_train, y_train), (X_test, y_test) = mnist.load_data()

# Normalize the data
X_train = X_train.reshape(-1, 28*28) / 255.0
X_test = X_test.reshape(-1, 28*28) / 255.0
```

### Build the Model

```
model = models.Sequential([
    layers.Dense(128, activation='relu', input_shape=(784,)),
    layers.Dropout(0.2),
    layers.Dense(64, activation='relu'),
    layers.Dropout(0.2),
    layers.Dense(10, activation='softmax')
```

])

Compile and Train:

```
model.compile(optimizer='adam',
loss='sparse_categorical_crossentropy',
metrics=['accuracy'])

model.fit(X_train, y_train, validation_data=(X_test, y_test), epochs=10, batch_size=32)
```

Evaluate and Test:

```
test_loss, test_acc = model.evaluate(X_test, y_test)

print("Test Accuracy:", test_acc)
```

**Visualize Results:** Plot sample predictions alongside the actual labels.

By completing this project, you'll gain hands-on experience with neural networks and a solid understanding of their real-world applications. In the next chapter, we'll explore techniques to optimize and deploy machine learning models.

# Chapter 06
## Machine Learning in Real-World Applications

Building a machine learning (ML) model is only the beginning of its journey. To deliver meaningful impact, the model must be implemented effectively, evaluated ethically, and scaled for real-world applications. This chapter delves into the practical aspects of deploying ML solutions, addressing ethical challenges, scaling for large datasets, debugging, and fostering collaboration. We'll conclude with a case study showcasing the power of ML for social good.

## Understanding the ML Lifecycle in Practice

The machine learning lifecycle extends beyond the training phase, encompassing deployment, monitoring, and iterative improvements. Understanding this process is crucial for bridging the gap between a prototype and a production-ready system.

### 1. From Prototyping to Production

**Prototype Stage:** Experimenting with different models and algorithms using a limited dataset.

- Tools: Jupyter Notebooks, local development environments.
- Focus: Rapid testing of ideas.

**Pre-production Stage:** Optimizing the model for real-world deployment.

- Convert notebooks into reproducible scripts.
- Validate the model using unseen datasets.

**Production Stage:** Deploying the model into a live environment where it interacts with users or systems.

- Use tools like Docker for containerization to ensure consistent behavior across environments.
- Monitor for performance and potential drift over time.

## 2. Continuous Monitoring

- Post-deployment, monitor:

✓ **Model Accuracy:** Is it still performing as expected?
✓ **Data Drift:** Are input data distributions changing over time?
✓ **Latency:** Does the system respond within acceptable time limits?

## Ethical Considerations in Machine Learning

The growing influence of ML makes ethics a critical concern. Poorly designed systems can perpetuate biases, invade privacy, or cause harm.

## 1. Addressing Bias

**Bias in Data:** Models learn from historical data, which may carry biases.

- Action Step: Audit datasets for imbalances and apply techniques like oversampling or reweighting.

**Bias in Algorithms:** Some algorithms may inherently favor certain outcomes.

➢ Action Step: Regularly test model outputs for fairness across different demographic groups.

## 2. Ensuring Fairness

➢ Use fairness metrics such as demographic parity or equal opportunity.
➢ Incorporate diverse perspectives during development to minimize blind spots.

## 3. Protecting Privacy

**Data Anonymization:** Remove personally identifiable information (PII).

**Federated Learning:** Train models on user data without storing it centrally.

**Regulatory Compliance:** Ensure adherence to laws lik GDPR (General Data Protection Regulation).

## Scaling ML Models for Large Data

Scaling models to handle large datasets efficiently is essential for real-world applications.

## 1. Introduction to Cloud Computing Platforms

➢ Cloud platforms like AWS, Google Cloud, and Azure provide tools for:

- **Storage:** Scalable storage solutions for big data.
- **Processing:** High-performance computing for training complex models.
- **Deployment:** Services like AWS SageMaker and Google AI Platform for hosting ML models.

### 2. Distributed Training

- **Why It's Needed:** Large datasets or complex models may exceed the capacity of a single machine.
- **Tools:**

- TensorFlow's Distributed Strategy API.
- Apache Spark for distributed data processing.

### 3. Efficient Data Pipelines

- Use tools like Apache Airflow or Kubeflow for building automated workflows:

- **ETL (Extract, Transform, Load):** Preprocess and move data efficiently.
- **Model Training:** Schedule and track experiments systematically.

## Debugging and Iterating Your Models

ML models often behave unpredictably, especially when applied to real-world scenarios. Debugging and refining them is a crucial part of the development cycle.

### 1. Identifying Common Pitfalls

- **Overfitting:** Model performs well on training data but poorly on new data.
- ✓ **Solution:** Use regularization techniques, increase training data, or simplify the model.
- **Underfitting:** Model fails to capture the underlying patterns in data.
- ✓ **Solution:** Increase model complexity or tune hyperparameters.

## 2. Debugging Strategies

- **Error Analysis:** Examine incorrectly predicted cases to identify patterns.
- **Visualization:** Plot confusion matrices, feature importances, and learning curves.

## 3. Iterative Refinement

- **Retrain on Updated Data:** Continuously integrate new data for improvement.
- **Hyperparameter Tuning:** Use tools like Grid Search or Bayesian Optimization to optimize parameters.

## Collaboration and Communication

Successful ML projects require collaboration across teams, ensuring that technical work aligns with broader goals.

## 1. Working with Cross-Functional Teams

- **Stakeholders:** Understand the business context and define success metrics collaboratively.
- **Domain Experts:** Incorporate domain-specific knowledge into feature engineering and evaluation.
- **Engineers:** Collaborate with software and DevOps teams for smooth deployment.

## 2. Documenting Work

- **Technical Documentation:** Record the rationale behind model choices and preprocessing steps.
- **Presenting Results:** Use clear visualizations and metrics to communicate findings to non-technical stakeholders.

## 3. Best Practices for Teamwork

- Establish regular check-ins and encourage knowledge sharing.
- Use tools like Git for version control and collaboration.

## Case Study: Using ML for Social Good

To highlight the real-world potential of ML, let's examine how machine learning is improving lives in underprivileged communities.

## Problem: Early Detection of Malnutrition

In developing regions, malnutrition in children often goes undetected due to resource limitations. ML can help by analyzing images and health records.

## Solution Overview

### Data Collection:

- Collect health indicators such as weight, height, and dietary habits.
- Use images of children to assess visible signs of malnutrition.

### Model Development:

- Use a CNN to analyze images for malnutrition indicators.
- Train regression models on health records to predict nutritional deficiencies.

### Deployment:

- Integrate the system into mobile apps used by health workers.
- Provide actionable insights such as nutritional advice or emergency alerts.

### Impact:

- Reduced the rate of undetected malnutrition cases by 40% in pilot regions.
- Enabled health workers to allocate resources more effectively.

This chapter demonstrates the transformative potential of machine learning when applied thoughtfully. As you progress, consider not only the technical challenges but also the societal implications of your work. The next chapter will dive into optimizing and deploying ML systems for seamless integration into everyday workflows.

# Part 4
# Innovating with Machine Learning

# Chapter 07
## Exploring Unsupervised Learning

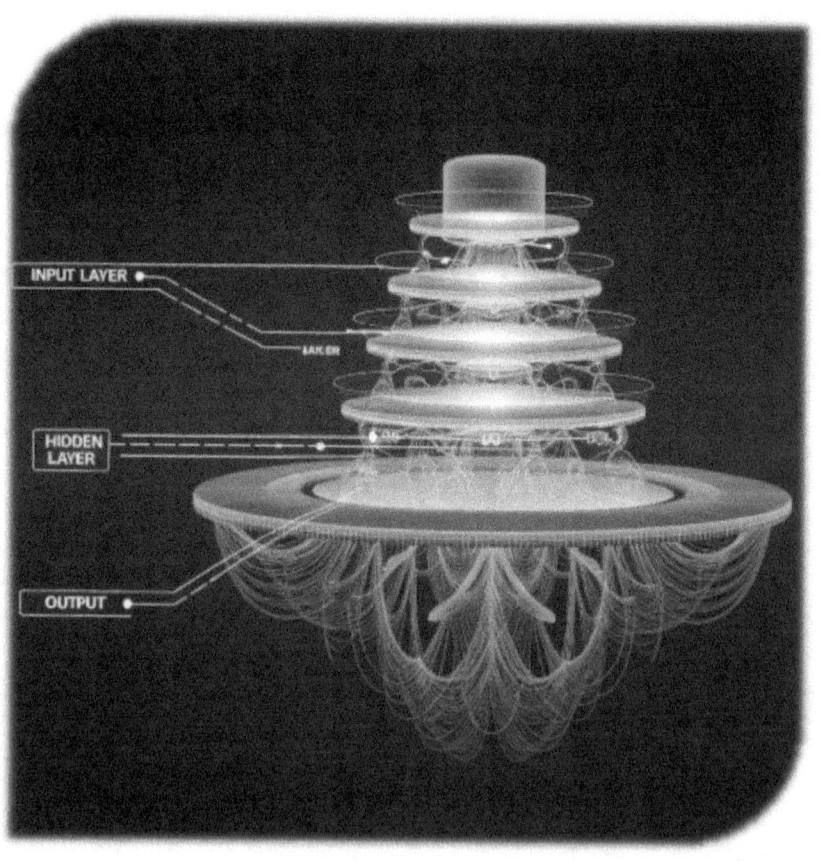

Unsupervised learning allows us to uncover hidden patterns in data without needing labeled outputs. This chapter will guide you through the core techniques of unsupervised learning, such as clustering and dimensionality reduction, and how to use these methods to extract meaningful insights from your data. Additionally, we will explore real-world use cases for unsupervised learning, demonstrate hands-on implementation in Python, and show how to combine unsupervised methods with supervised models to enhance your projects.

## Clustering Techniques

Clustering is one of the most widely used techniques in unsupervised learning, helping us group similar data points based on certain characteristics. Understanding clustering algorithms will allow you to find patterns in data and create meaningful segments for analysis.

### 1. K-Means Clustering

#### What is K-Means?
K-means is an algorithm that divides data into K distinct clusters. The algorithm works by:

1. Initializing K centroids randomly.
2. Assigning each data point to the nearest centroid.
3. Recomputing the centroids based on the mean of the points assigned to each cluster.

Repeating the process until convergence.

## Steps to Implement K-Means in Python:

- Import necessary libraries (`scikit-learn`, `matplotlib`).
- Load your dataset (e.g., from a CSV or a data frame).
- Apply K-means algorithm and define the number of clusters.
- Visualize clusters using scatter plots.

## When to Use K-Means:

- Ideal for datasets where clusters are spherical and evenly sized.
- Works well when you have a rough idea of the number of clusters you want to detect.

# 2. Hierarchical Clustering

## What is Hierarchical Clustering?
Hierarchical clustering builds a tree-like structure called a dendrogram that represents the merging or splitting of clusters. It can be either:

- **Agglomerative:** Start with individual points and merge them step-by-step.
- **Divisive:** Start with all points in one cluster and split them iteratively.

## Steps to Implement Hierarchical Clustering:

1. Calculate the distance between points using a linkage method (e.g., Ward's method).
2. Build the dendrogram.

3. Cut the dendrogram at the desired level to form clusters.

## When to Use Hierarchical Clustering:

- When you want to visualize the data's structure in a tree-like diagram.
- When you don't know how many clusters to expect.

## Dimensionality Reduction

High-dimensional datasets can lead to overfitting, computational inefficiency, and difficulty in visualizing the data. Dimensionality reduction techniques help reduce the number of features (variables) while retaining the essential patterns in the data.

## 1. Principal Component Analysis (PCA)

### What is PCA?

PCA is a technique used to reduce the number of dimensions in a dataset while maintaining most of the variation in the data. It transforms the original features into a smaller set of "principal components" that explain the most variance in the dataset.

### Steps to Implement PCA:

1. Standardize the dataset (important for PCA).
2. Calculate the covariance matrix to understand how variables correlate.
3. Compute eigenvectors and eigenvalues.
4. Choose the top K components and transform the dataset accordingly.

## When to Use PCA:

- ➤ When you need to reduce the complexity of your dataset.
- ➤ When dealing with datasets with many correlated features.

## 2. t-Distributed Stochastic Neighbor Embedding (t-SNE)

### What is t-SNE?

t-SNE is a non-linear dimensionality reduction technique primarily used for visualizing high-dimensional data. It preserves local structure by focusing on minimizing the divergence between probability distributions in high- and low-dimensional spaces.

### Steps to Implement t-SNE:

1. Import `TSNE` from `sklearn.manifold`.
2. Apply t-SNE to the dataset to reduce dimensions to 2 or 3.
3. Visualize the result with a scatter plot.

### When to Use t-SNE:

- ➤ Ideal for data visualization, especially when you want to display high-dimensional data in 2D or 3D.
- ➤ Useful in exploring clusters and outliers in datasets.

## Use Cases for Unsupervised Learning

Unsupervised learning is used across various domains to uncover insights, create new features, and identify hidden patterns in data. Below are some real-world use cases where unsupervised learning excels.

## 1. Anomaly Detection

### What is Anomaly Detection?
Anomaly detection is used to identify unusual patterns that do not conform to expected behavior, such as fraud detection in financial transactions or identifying defects in manufacturing.

### How Unsupervised Learning Helps:

- Algorithms like K-means or isolation forests can detect outliers in data without labeled examples.
- In fraud detection, anomalies like unusual spending patterns are flagged for further investigation.

## 2. Customer Segmentation

### What is Customer Segmentation?
Unsupervised learning can be used to segment customers based on purchasing behavior, demographics, or engagement. This helps businesses personalize marketing strategies and optimize product offerings.

### How Unsupervised Learning Helps:

- Techniques like K-means clustering can group customers based on similarities in their purchasing behavior.
- PCA can help reduce the complexity of customer data while preserving essential patterns for segmentation.

# 3. Market Basket Analysis

## What is Market Basket Analysis?

This technique analyzes customer purchase behavior to identify associations between different products. It's commonly used in retail to recommend products that are frequently bought together.

## How Unsupervised Learning Helps:

- Apriori or frequent pattern mining algorithms can detect product combinations that appear together frequently.
- This information can be used to design better store layouts or online recommendations.

## Building and Visualizing Clusters in Python

Let's apply what we've learned by clustering a dataset in Python. We will use the `scikit-learn` library to build clusters and visualize them using `matplotlib`.

## Step-by-Step Implementation

### Load Data:

- Import the dataset (e.g., `iris.csv` or `customer_data.csv`).
- Check for any missing data and handle it appropriately.

### Preprocess Data:

> Standardize the dataset using `StandardScaler` to ensure that all features are on a similar scale.

## Build Clusters:

> Apply K-means or hierarchical clustering.
> Set the number of clusters (K) for K-means or use a dendrogram for hierarchical clustering.

## Visualize the Clusters:

> Use `matplotlib` to plot the data points and color them according to the clusters.
> For high-dimensional data, apply PCA or t-SNE to reduce dimensions for visualization.

## Combining Unsupervised Learning with Supervised Models

Often, combining unsupervised learning with supervised learning can yield powerful models. Unsupervised learning can be used to preprocess data, extract features, or cluster data, which is then fed into supervised learning algorithms for predictions.

### 1. Feature Engineering

> Use unsupervised learning to find hidden patterns and use the results as new features for a supervised learning model.

### 2. Semi-supervised Learning

> Combine labeled data with large amounts of unlabeled data. Use unsupervised learning to label the unlabeled data and enhance the performance of the supervised model.

## Project: Segment Customers Based on Purchasing Behavior

Now, let's create a project to apply unsupervised learning in a practical scenario.

### Project Overview:

We will segment customers based on their purchasing behavior using clustering techniques.

### Steps:

### Collect Data:

> Use data from an online store that includes purchase history and demographic details.

### Preprocess Data:

> Handle missing values, normalize the data, and encode categorical features.

### Clustering:

> Apply K-means to group customers into segments based on their behavior (e.g., frequent buyers, one-time shoppers, etc.).

## Evaluation:

➢ Analyze the clusters to derive actionable insights (e.g., targeting marketing campaigns to frequent buyers).

By mastering unsupervised learning techniques, you'll be equipped to uncover patterns, reduce complexity, and gain deeper insights into your data. This chapter has laid the groundwork for integrating unsupervised learning into your machine learning projects. In the next chapter, we will explore reinforcement learning, a unique approach that enables machines to learn from interactions with their environment.

# Chapter 08

## Reinforcement Learning and Future Trends

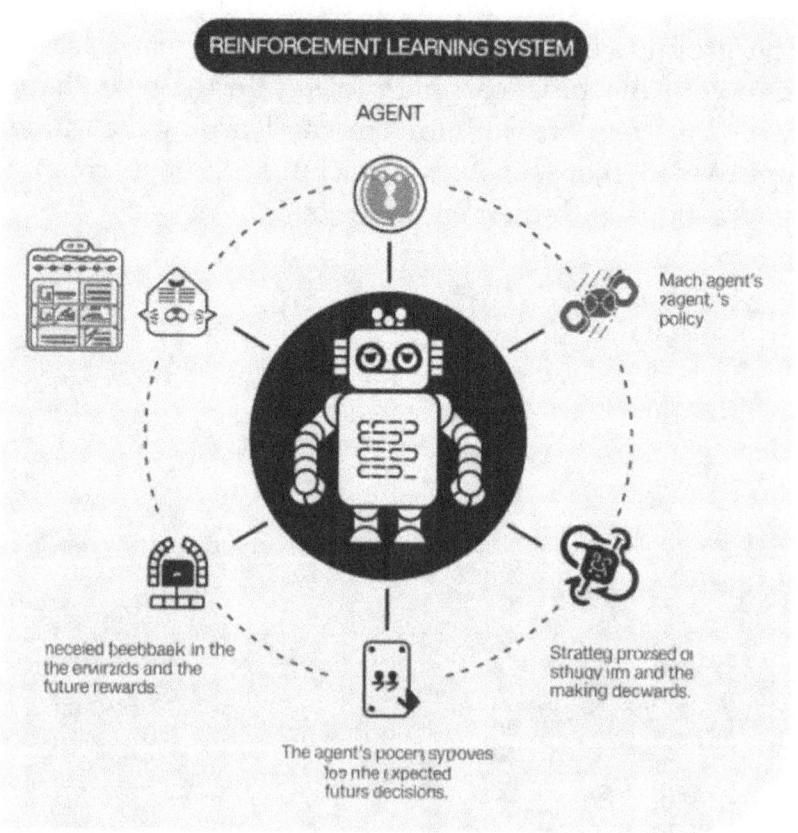

Reinforcement learning (RL) represents an exciting frontier in machine learning, where an agent learns how to make decisions by interacting with an environment and receiving feedback in the form of rewards. Unlike traditional supervised learning, where the model learns from labeled data, RL enables systems to learn from trial and error. In this chapter, we will explore the fundamentals of reinforcement learning, its real-world applications, and dive into cutting-edge areas like deep reinforcement learning. We will also discuss the future trends in machine learning and AI, touching upon generative models, edge computing, and quantum machine learning.

## What is Reinforcement Learning?

Reinforcement learning is a type of machine learning where an agent learns to take actions in an environment to maximize a cumulative reward. The agent interacts with the environment, makes decisions, and receives feedback based on those decisions. Over time, the agent learns the best strategies or policies to achieve its goals.

## 1. Basics of Agents, Environments, and Reward Systems

### The Agent:
An agent is the learner or decision maker that interacts with the environment. It can be anything from a robot trying to navigate through space, to a software program optimizing strategies.

### The Environment:
The environment is everything the agent interacts with.

It could be a virtual world, a physical system, or a process the agent controls. The environment provides feedback in the form of states, actions, and rewards.

### The Reward System:

Rewards are signals given to the agent to evaluate the quality of the actions it takes. Positive rewards encourage the agent to repeat successful actions, while negative rewards (penalties) discourage undesirable behaviors. The goal is for the agent to maximize its cumulative reward over time.

- **State:** A snapshot of the environment at a given time.
- **Action:** A decision the agent makes based on its current state.
- **Reward:** Feedback from the environment after the agent takes an action.

## 2. Key Elements in Reinforcement Learning

- **Policy:** The strategy the agent uses to determine its actions based on the current state. It could be deterministic or probabilistic.
- **Value Function:** A function that estimates the long-term reward the agent can expect from a given state or action.
- **Q-value:** A function that helps the agent decide which action to take by evaluating the expected reward for each action in a given state.

### Real-World Applications

Reinforcement learning is being widely adopted in various industries and is revolutionizing many domains, from gaming to robotics. Below are some of the most exciting real-world applications of RL.

## 1. Gaming

### DeepMind's AlphaGo:
AlphaGo, created by DeepMind, is a prime example of RL applied to gaming. It mastered the game of Go, which is considered much more complex than chess, by using a combination of deep learning and reinforcement learning. AlphaGo learned through self-play, where it played millions of games against itself to improve its strategies.

### Atari Games:
OpenAI's reinforcement learning agents have been trained to play Atari games. The agent starts with no knowledge of the game but learns optimal strategies purely through trial and error.

## 2. Robotics

### Autonomous Robots:
RL is widely used in robotics to enable robots to learn complex tasks such as walking, object manipulation, and autonomous navigation. By rewarding robots for taking correct actions, they can gradually improve their ability to perform physical tasks without human intervention.

### Industrial Automation:
RL is also applied to optimize robotic systems in

manufacturing and logistics, where robots learn to perform tasks like assembling components, sorting packages, or navigating warehouses more efficiently.

### 3. Autonomous Systems

### Self-Driving Cars:

Autonomous vehicles use RL to learn how to make decisions such as when to accelerate, brake, and turn. The vehicle's "agent" learns by receiving feedback based on its actions, such as staying within lanes, avoiding collisions, and following traffic rules.

### Drones:

Drones utilize RL to perform tasks like flying through complex environments, avoiding obstacles, and performing aerial inspections or deliveries.

## Deep Reinforcement Learning

Deep Reinforcement Learning (DRL) is the combination of reinforcement learning and deep learning. By incorporating deep neural networks, DRL enables agents to learn complex representations of their environment and make decisions based on high-dimensional data like images or video streams.

### 1. Introduction to Q-Learning

### What is Q-Learning?

Q-learning is one of the most well-known algorithms in reinforcement learning. It is a model-free algorithm that learns the value of an action taken in a given state by updating its Q-values based on the reward received.

- **Q-Table:** A table where each entry contains a Q-value representing the expected future reward for a given state-action pair.
- **Action Selection:** The agent uses the Q-table to choose the action that maximizes the expected future reward.

## The Q-Learning Update Rule:
The agent updates its Q-values using the formula:

- $$Q(s_t, a_t) = Q(s_t, a_t) + \alpha [r_t + \gamma \max_a Q(s_{t+1}, a) - Q(s_t, a_t)]$$

Where:

- $\alpha$ is the learning rate.
- $r_t$ is the reward at time $t$.
- $\gamma$ is the discount factor.

## 2. Policy Gradients

### What are Policy Gradients?
Policy gradients are a class of RL algorithms that directly optimize the policy by adjusting the policy parameters to increase the expected cumulative reward. Unlike Q-learning, which is based on value estimation, policy gradients focus on directly modifying the policy.

> **Advantage of Policy Gradients:**
> They are particularly useful in environments with continuous or high-dimensional action spaces, such as robotics or self-driving cars.

## Frameworks for Reinforcement Learning

When working with reinforcement learning, developers often use specialized frameworks to streamline the implementation process. Below are some of the most widely used RL frameworks.

## 1. OpenAI Gym

### What is OpenAI Gym?
OpenAI Gym is a popular toolkit that provides a wide range of environments for developing and testing RL algorithms. It supports environments for games, robotics, and control systems, allowing you to build and train RL agents in diverse scenarios.

### Features of OpenAI Gym:

> Predefined environments with standard interfaces.
> Easy to integrate with other RL libraries such as TensorFlow or PyTorch.
> Allows users to create custom environments.

## 2. RLlib

### What is RLlib?
RLlib, developed by Ray, is a scalable reinforcement learning library built on top of Ray, a distributed

computing framework. RLlib supports multiple RL algorithms and is optimized for use in both research and production environments.

## Key Features of RLlib:

- ➢ Scalable to large environments with distributed training.
- ➢ Supports deep learning frameworks such as TensorFlow and PyTorch.
- ➢ Easy integration into production systems.

### 3. TensorFlow Agents

- ➢ **What is TensorFlow Agents?**
  TensorFlow Agents is an RL library built on TensorFlow that provides modular components for building RL agents. It is designed for research and allows easy experimentation with different algorithms.

### The Future of ML and AI

Machine learning and artificial intelligence are evolving rapidly, and several trends are shaping the future of this technology. From generative AI to quantum computing, the possibilities seem limitless.

### 1. Generative AI

**What is Generative AI?**
Generative AI refers to systems that can create new content, such as text, images, or music, based on learned patterns from data. GPT models and GANs (Generative

Adversarial Networks) are prominent examples of generative AI technologies.

## Applications:
Generative AI is revolutionizing industries like entertainment (e.g., content creation), healthcare (e.g., drug discovery), and art (e.g., AI-generated artwork).

# 2. Edge Computing

## What is Edge Computing?
Edge computing refers to processing data closer to where it is generated, rather than sending it to centralized servers. In the context of AI, edge computing enables faster decision-making in real-time applications like autonomous driving and smart devices.

## Impact on AI:
Edge computing allows AI models to run more efficiently on devices with limited resources, enhancing privacy and reducing latency.

# 3. Quantum Machine Learning

## What is Quantum Machine Learning?
Quantum machine learning combines quantum computing with traditional machine learning to create algorithms that can process complex data at unprecedented speeds.

## Potential Impact:
Quantum ML could revolutionize industries that rely on

complex simulations, like chemistry, cryptography, and optimization problems.

## Innovative Project Idea: Train a Virtual Agent

Now, let's take the principles of reinforcement learning and put them into practice with a hands-on project idea. You will train a virtual agent to perform a simple task in a simulated environment.

### Project Overview:

- **Objective:** Create a virtual agent that can learn to navigate a maze and reach a goal using reinforcement learning.
- **Tools Needed:** Python, OpenAI Gym, TensorFlow, or PyTorch.

### Steps:

**Setup Environment:**
Use OpenAI Gym's maze environment to simulate the agent's task.

**Define Rewards:**
Assign positive rewards for reaching the goal and negative rewards for hitting obstacles.

**Train the Agent:**
Use Q-learning or a deep reinforcement learning model to train the agent.

**Evaluate:**
Test how well the agent performs and refine the model based on performance.

Reinforcement learning is an exciting and rapidly advancing field with many real-world applications. As AI continues to evolve, reinforcement learning will be a key driver in creating smarter systems that learn from experience. The future is bright, and by mastering RL, you'll be prepared to take on the next generation of AI challenges.

# Conclusion

# Embracing the Journey

Congratulations! You've just taken a monumental step toward mastering one of the most exciting and transformative fields of technology today—Machine Learning. While this book has provided you with the foundational knowledge and tools to get started, it's only the beginning. The world of ML is vast, and the potential to innovate and create is endless. Let's reflect on how far you've come, and explore how you can continue to grow, make an impact, and embrace the future of machine learning.

## Reflecting on Your Growth

As we close this guidebook, take a moment to appreciate how much you've learned about machine learning (ML) and the world of smart machines. From understanding the basics of algorithms and data to building and fine-tuning your own models, you've gained practical knowledge and skills that will serve as the foundation for further exploration. You're no longer just a beginner in the field; you're well on your way to becoming an innovator and problem solver.

## Key Milestones:

- You've learned the foundational principles of ML, such as the differences between supervised, unsupervised, and reinforcement learning.
- You've had hands-on experiences, like training your first ML model, working with data, and understanding the nuances of algorithms.

> You've explored cutting-edge areas, including deep reinforcement learning and neural networks, and have already built a small but impactful project.

This journey represents a crucial turning point in your education. Remember, while technical knowledge is important, it's your curiosity, persistence, and creativity that will help you stand out. ML is more than just coding and mathematics—it's about solving real-world problems and making meaningful contributions.

## Continuing the Learning Journey

Machine learning is a constantly evolving field, with new research, tools, and technologies emerging regularly. To stay ahead of the curve, you'll need to continue your education. Fortunately, the ML community is vibrant and filled with resources that can help you grow.

## Resources for Further Exploration:

### Books:

> *"Hands-On Machine Learning with Scikit-Learn, Keras, and TensorFlow"* by Aurélien Géron is an excellent follow-up for expanding your practical skills.
> *"Deep Learning"* by Ian Goodfellow provides an in-depth exploration of deep learning models.

### Online Courses:

> **Coursera's Machine Learning Course** by Andrew Ng is a timeless resource for beginners.

- **fast.ai** offers free courses that focus on practical ML and deep learning.

## Communities and Forums:

- **Kaggle** is a platform where you can participate in competitions and collaborate with other data scientists.
- **Stack Overflow** and **Reddit's Machine Learning Subreddit** provide spaces for asking questions and engaging with experienced professionals.

## Conferences and Events:

Attending events like *NeurIPS*, *ICML*, or *CVPR* will expose you to the latest research and innovations in ML. Many events offer workshops and tutorials perfect for deepening your knowledge.

## Tips for Innovators

Machine learning knowledge is a powerful tool, but how you apply it is what will set you apart. Whether you are a student, an entrepreneur, or an innovator, here are some tips for using ML to solve unique problems and create groundbreaking solutions:

### 1. Start with a Problem

- Rather than diving into algorithms or models first, focus on a problem that interests you. Whether it's predicting disease outcomes, automating tasks, or optimizing energy usage, find a real-world issue that you're passionate about solving. ML is best applied to problems that can benefit from data-driven insights.

## 2. Build Iteratively

➤ Don't try to build the perfect model on your first try. Start small, build a prototype, and iterate. Refining your models based on real-world feedback is key to success in ML.

## 3. Think Cross-Disciplinary

➤ ML has applications across virtually every field. Whether you're in healthcare, finance, agriculture, or entertainment, the possibilities are endless. Collaborate with experts in other fields to design solutions that leverage both domain knowledge and ML techniques.

## 4. Embrace Failure

➤ Machine learning is all about experimentation. Not every model will succeed, and not every approach will lead to breakthrough results. Embrace failure as a learning opportunity and use it to refine your understanding and approach.

## Inspiration for Future Projects

Your ML journey doesn't end here—it's just the beginning. The skills you've acquired will empower you to tackle some of the most pressing challenges facing our world today. To inspire your next steps, consider these project ideas that blend creativity with technical expertise:

## Project Ideas:

## Smart Healthcare:

- ➢ Build a model to predict disease outbreaks or analyze medical images for early detection of conditions such as cancer. The possibilities for using ML to improve healthcare are vast and transformative.

## Sustainability Projects:

- ➢ Use ML to optimize energy consumption, predict environmental changes, or develop more efficient recycling systems. With growing concerns about climate change, ML applications in sustainability will continue to be in high demand.

## Financial Technology:

- ➢ Develop predictive models for stock market trends or create algorithms for fraud detection. The financial sector is ripe for innovation through ML, and the possibilities for improving efficiency and security are abundant.

## Personalized Learning:

- ➢ Create a learning platform that tailors educational content based on students' progress and preferences. ML can revolutionize education by providing customized learning experiences that improve student engagement and outcomes.

## Your Role in Shaping the Future

As you continue to grow and apply your knowledge of machine learning, remember that the impact of your ideas is limitless. You are part of a new generation of thinkers and innovators who are reshaping industries, solving complex problems, and advancing society.

### 1. Innovators as Problem Solvers

- Machine learning has the power to solve real-world problems in areas like healthcare, climate change, education, and more. As you move forward, think of yourself not just as a developer or researcher, but as a problem solver who can use ML to make the world a better place.

### 2. Ethical Considerations

- As you innovate, keep ethical considerations in mind. The decisions you make about data, models, and applications will have long-lasting implications. Be mindful of issues like bias in data, privacy concerns, and the potential for AI to be used in harmful ways. The responsibility lies with creators to ensure technology is developed for the betterment of society.

### 3. Continuous Growth

- The field of ML is fast-paced and ever-evolving. Your journey will require ongoing learning and adaptation.

Stay curious, stay motivated, and never stop experimenting. The future is full of new opportunities and challenges, and you are equipped to shape it.

## Embrace the Future

Machine learning is a transformative technology with the potential to change how we live, work, and interact with the world. By building on the knowledge and skills you've gained in this book, you're ready to explore its endless possibilities. Embrace the journey ahead, and remember—your ideas, your innovations, and your passion for learning will help shape the future of technology.

So, what's next for you? The world is waiting for your creativity and expertise. Keep learning, keep building, and keep imagining the impossible. The future is in your hands!

www.ingramcontent.com/pod-product-compliance
Lightning Source LLC
Chambersburg PA
CBHW071653240526
45469CB00023B/2312